The Exhibition Dorking Chicken
Hints to Exhibitors and Poultry Fanciers of the Dorking Fowl

by Thomas Coke Burnell

with an introduction by Jackson Chambers

This work contains material that was originally published in 1875.

This publication is within the Public Domain.

This edition is reprinted for educational purposes and in accordance with all applicable Federal Laws.

Introduction Copyright 2017 by Jackson Chambers

Self Reliance Books

Get more historic titles on animal and stock breeding, gardening and old fashioned skills by visiting us at:

http://selfreliancebooks.blogspot.com/

Introduction

I am pleased to present yet another title in the "Chicken Breeds" series.

This volume is entitled "The Exhibition Dorking". It was originally published in 1875 by Thomas Coke Burnell of England.

Included are important historic details on the breeding and exhibiting of the legendary Dorking Fowl.

The work is in the Public Domain and is re-printed here in accordance with Federal Laws.

Though this work is a century old it contains much information on poultry that is still pertinent today.

As with all reprinted books of this age that are intended to perfectly reproduce the original edition, considerable pains and effort had to be undertaken to correct fading and sometimes outright damage to existing proofs of this title. At times, this task is quite monumental, requiring an almost total "rebuilding" of some pages from digital proofs of multiple copies. Despite this, imperfections still sometimes exist in the final proof and may detract from the visual appearance of the text.

I hope you enjoy reading this book as much as I enjoyed making it available to readers again.

Jackson Chambers

PREFACE.

THE following "hints" appeared first in the *Journal of Horticulture*; at the request of several amateurs. I now reprint them, trusting they may be of some use to beginners in a pursuit which to me has such a fascination.

T. C. BURNELL.

Stratton, Micheldever, Hants,
May, 1875.

THE EXHIBITION DORKING.

CHAPTER I.

THE ADVANTAGES OF DORKINGS.

DORKINGS have always stood well in the estimation of the poultry-loving public, even before their fine size and comely shape were brought prominently into notice by public exhibitions. So far back as 1853 birds of this breed realized high prices, and in that year the Rev. — Boyes sold his prize pen at Hitchin for £50, and the Rev. S. Donne lost his prize birds at the Midland Counties Show, although protected by twenty guineas. Other instances of equally high prices are not wanting. In 1867 Lady Holmesdale's yard of Dorkings, which had been under the management of Mr. John Martin, realized over £400; and the prices given at this sale for individual birds would almost exceed belief. To come down to the present day, I will only instance my pen of chickens at Oxford last year, which were claimed at the catalogue price of twenty guineas, after winning Prince Leopold's cup; and my first-prize cock at the Crystal Palace Show this year sold for the same price; and I am more than ever convinced that £25 would not now be sufficient to protect a single cock, were he the best of the year and a likely stock bird.

With such a ready sale for good birds at high prices in prospect, surely a fair field for surplus energy is open to the dweller in the country. For my part I know what it is to have had to give up my profession through delicate health; and I am sure there are many like myself, who would hail with delight an occupation which gains upon one with success, and which combines with fresh air, an occupied mind, a fair amount of excitement, and a good prospect of substantial profits. That the latter is no chimera I can positively state from experience; and lest anyone should be deterred by a want of knowledge, I will add that four years ago I knew as little about exhibition poultry as the most ignorant of my readers. In order that I may not be considered egotistical I will here say that the views I shall express on Dorkings are not given solely on my own authority, but only after careful study of previous works on the subject, and numerous pleasant chats with some of the most noted breeders of the day. Although Dorkings are my particular fancy, I will not ask anyone to suppose that I think them suited to every exhibitor and every situation. To anyone who has a moderate grass run and a desire for a plentiful supply of very superior chickens for the table, with a fair amount of eggs, I believe they are unrivalled; but for damp back yards and other confined spaces they are altogether unsuited.

It is often said Dorkings are only adapted to a gravelly or chalky soil; but this is a mistake, or neither Mrs. Arkwright nor Admiral Hornby would have been so successful, their poultry runs being situated on a stiff clay soil. I quote these instances in order that no amateur may be deterred, but at the same time think that anyone situated in a damp locality would be better suited with the yellow-legged breeds, though not so well adapted for the table.

As to the Dorking being tender, I can only say that last year I reared over a hundred Dorking chickens and only lost one; but this is too fortunate an average to take for any breed.

I have always found Dorkings fair layers of large-sized eggs; and so precocious are the pullets, that it is one of the principal difficulties of the exhibitor in this breed to keep them from laying and too early maturity.

Of the different varieties, the Cuckoos are the best layers; but then, they do not reach to the size of the coloured birds. For farmyard and useful purposes I prefer the Coloured and Cuckoos; while, if it is desired to combine a really useful fowl with an attractive form, I think the Whites and Silvers are unequalled.

To the would-be exhibitor Dorkings possess one very great advantage over every other breed, and an advantage, too, that cannot be lightly estimated—they are within the power of every honest fancier to prepare for show; there are no vulture hocks to pluck and curl, no hackles to pull; neither have they to be kept shut up in the dark to bring out their colours. Anyone who knows what is constantly done and exposed in some breeds will not estimate these advantages too lightly; and it is for this reason more than any other that I venture to recommend Dorking fowls to anyone wishing to become a poultry fancier.

CHAPTER II.

BUYING FOWLS AND EGGS.

I WILL now suppose that some one of my readers has determined to take up Dorking fowls for exhibition, or, perhaps, is only desirous of improving his present stock of poultry by the introduction of fresh blood of the best strains. The question will be, Which is the best way to proceed? In my own case I attended the nearest poultry exhibition with the intention of buying one or two of the best birds for, as I expected, a pound or two. You may imagine my disgust on finding all the prize birds priced at £100, and all the rest at nearly equally high prices. However, I presently came to the "selling classes," where the price of the competing specimens is usually limited to 30s.; and seeing that the second-prize bird was a Dorking cock entered at that price, and, by the catalogue, eight months old, I rushed off to the Secretary's office, and after a good deal of pushing and struggling secured him at that price; also a pair of highly-commended hens at the same figure, and returned home very well satisfied with my day's work.

I was so pleased with my new purchases that I invited the poultryman from the neighbouring farm to be present when they were unpacked, and busied myself immensely with wood-work and wire netting to have a separate place for them on arrival, so that they might not be contaminated by low-born company. The birds appeared in good time, and were let out and fed; and I confess that they did not look quite so well to me on the ground as when I saw them in a show-pen: and I may as well here add that I have found this opinion strengthened by experience; and my readers may take it as a rule that seeing birds at an exhibition is very deceptive, and that if the pens are placed high up it makes them look half as large again. However, to my story: the man caught hold of the cock, and passing his hand down his breastbone, remarked that it was so crooked "that you could put your fist into it," and asked me to look at the length and sharpness of his spurs, and the scales on his legs, and said he was certain that the bird was at the very least five years old. I strenuously denied this, as the catalogue said he was eight months old. The hens, however, seemed to please him, as he remarked they were very large; and so they were left.

I could not keep long away from my new purchase, and on returning to the pen in an hour or two I found the cock's head all covered with blood, and one of the hens' beak in the same state. This rather astonished me, and I thought the cock and hen must have been fighting, which I considered very ungallant on the part of the old Dorking cock, who I had always heard styled "the pattern of an English gentleman." I watched them for a little while, and soon saw the hen go up to the cock very affectionately and commence to peck his comb, which was already streaming with blood; and to my astonishment the latter seemed to enjoy it; but I now thought it high time to interfere, before the cannibal hen should have quite eaten his comb away. To make a long story short, the cock proved old and useless, one of the hens was an inveterate comb-eater, while the other laid shell-less eggs, the extreme value of the three being about 4s. to make into soup, which was their ultimate destination.

I could give more instances of disappointment from buying birds in a "selling class" were it necessary, but I shall only mention this one as a sample of the others, so that intending purchasers in a "selling class" may draw their own inferences. I do not for an instant deny that bargains are occasionally to be picked up in "selling classes," but only by good judges who are able to appreciate the merits of birds which some amateur does not know the value of; but I am more than ever convinced that beginners had better steer clear of them. In trying, then, to solve the question, Where are good birds to be obtained? I will not attempt to give advice to old fanciers, most of whom are well able to teach me, but simply, if possible, to give a few hints in a crude form to assist beginners.

I recommend anyone really ignorant of fancy points and all that is required in a good bird, and who has not sufficient confidence in his own opinion, nor time to attend an auction, to apply to one of the well-known dealers, who, if he asks him a good price, will give him a good bird; or else to write to one of the numerous successful exhibitors of the day, stating his wants and the amount he intends to give, leaving all details to the exhibitor, and trusting him to send him the value for his money. I have the pleasure of the acquaintance of nearly all the Dorking exhibitors, and am convinced that not one would take an advantage of a beginner were he to evince confidence; but when anyone writes pretending to be a good judge, the exhibitor will often expect him to find out the defects for himself.

There are many who do not like to buy birds without previously seeing them. I would recommend such to buy their birds "on approval," by which is commonly meant that if the fowls are not approved of they may be returned at once, the intending purchaser, of course, paying all carriage both going and returning, and also being answerable that the birds reach home in safety. Of course any special agreement can be made that is thought necessary, but buyers are invariably expected to send the money before the birds are sent off. An honest purchaser should have no objection to this, as it will be readily understood that in these days of sharp practice vendors must be on their guard, as many apply to have birds sent them on approval who have not the least intention of either paying for or returning them.

The next question will be, At what time of year are we most likely to procure birds at a moderate price? This will depend upon what it is we require. A really good bird, and one likely to win at good shows, has always a certain value, and I should look with suspicion on any advertiser who offers such at a very low price; but inferior birds, "wasters," as they are called, are much more plentiful at one time of year than another. I should recommend anyone who keeps four or five cocks running together in one yard, and who wishes for a change of blood of the best strains, to apply to one of the large breeders and exhibitors about June or July in each year, when they must have a large number of chickens from ten to fifteen weeks old, and when they would generally be only too glad to get rid of half-a-dozen young cockerels showing slight defects for exhibition, but equally good as their best for the purpose required, at from 10s. to 15s. each. Necessarily, for a single bird they would charge rather more. One great advantage of procuring cockerel chickens of this age is, that they will not attempt to fight the old birds if introduced into a strange yard; while they will grow up with your own chickens, and you will escape all the fighting and destruction which is the inevitable result of introducing a full-grown cock into a new yard. This object may also be attained by purchasing sittings of eggs; but in doing so too great care cannot be expended in ascertaining in the first place whether the advertiser really has good stock, and secondly, whether, if he has, he will let you have the eggs from them. There are many complaints of bought eggs not hatching, but anyone who knows what a little is required to spoil a clutch of eggs will not always attribute failure to the roguery of the vendor, though I am sorry to express my belief that the latter is sometimes the case.

Fanciers always purchase eggs at their own risk, and it cannot be fair to make the vendor answerable for them after leaving his hands.

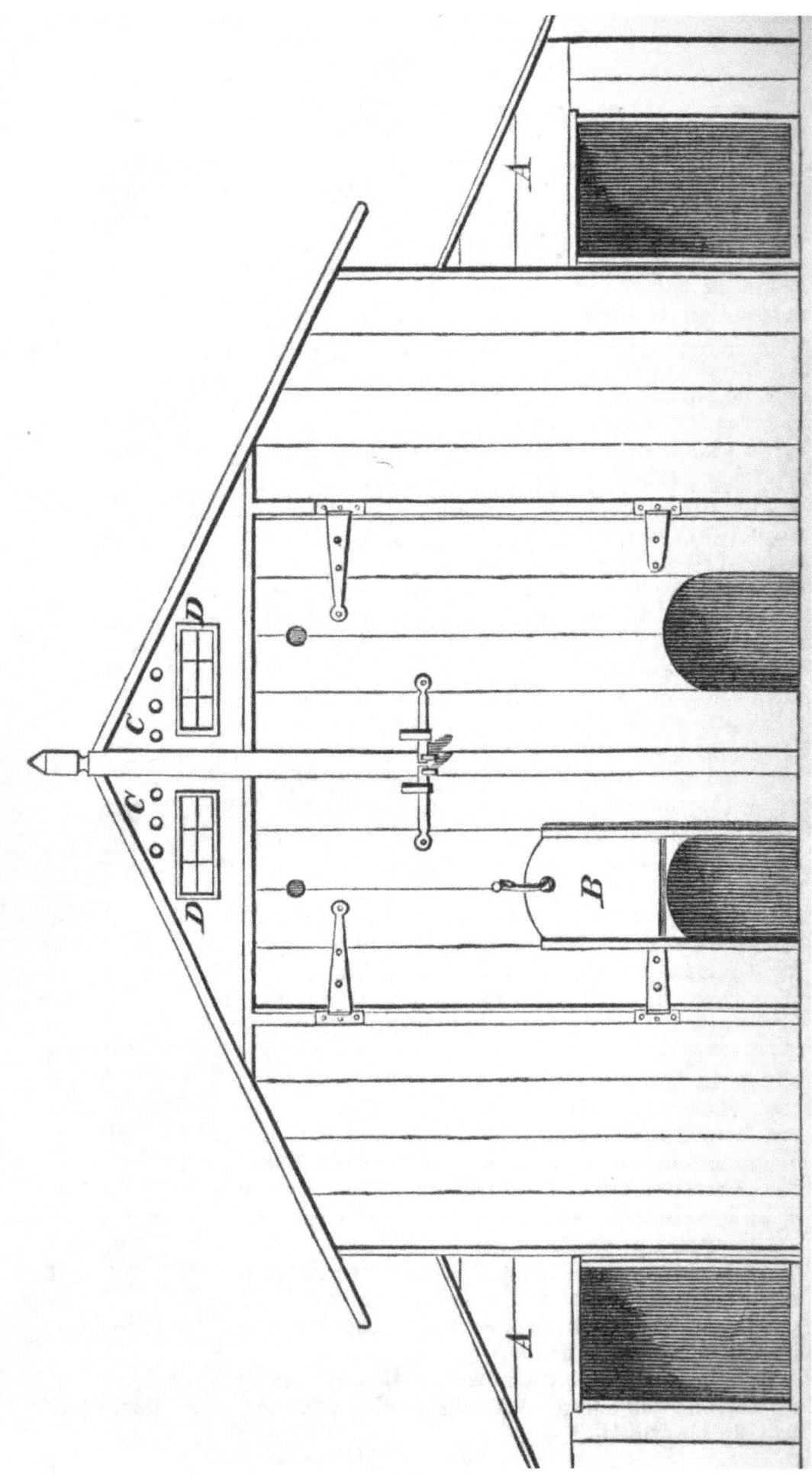

CHAPTER III.

POULTRY HOUSES AND RUNS.

Like any other pursuit, success with prize poultry is only to be attained by paying great attention to numerous small details. Anyone expecting to breed and rear valuable birds without any trouble will be most certainly disappointed, while at the same time I think there is no stock which will so well repay any labour that may be expended upon it.

It is often said, "How lucky So-and-so is;" but depend upon it there is very little luck about it, or we should never see the same exhibitors so continually coming in at the top of the prize list. Occasionally, either when birds get into a dark corner or the judges are overworked, a slight error may occur, but such is sure to be rectified at the next show.

It may seem a small matter where and on what our birds roost, but such will not be found to be the case. No very elaborate place is required, and in most stable-yards there is some shed or outbuilding which may readily be converted into a poultry-house, provided only that it is dry and light. If the floor be of brick or stone such had better be removed or covered over with 3 or 4 inches of earth or sand, as cold floors are generally thought to cause disease. One of the most necessary points is that the house should be well ventilated without being draughty. Fresh air cannot hurt anything, but a chilly draught blowing right across the perch is very different. The best way to provide fresh air without a draught is to have two or three good-sized openings in the top of the south side of the house, which must not be too low, and for the perches to be placed some way below the ventilators.

If the house be dark the fowls will not enter if they can possibly help it, while if we have a sliding window it may be advantageously taken out in summer to allow the entrance of fresh air.

The inside of the house should be limewashed occasionally —a simple and inexpensive operation, which will go a long way towards preventing disease; while the floor and perches should be cleaned at least twice a-week (better still every day), and chloride of lime, carbolic acid, or some other disinfectant sprinkled about. McDougall's disinfecting powder is very convenient for this purpose, also for sprinkling the nests with; and I believe if these precautions were generally adopted we should hear less of vermin in fowl houses—at all events, I *never* have such a thing in my own.

As Dorkings are such large, massive fowls, the perch should not be placed too high from the ground, otherwise in descending from roost of a morning the birds are apt to injure the ball of the foot, a fertile cause of the inflammation which is called bumble-foot, and which I will allude to under the head of Diseases. The perches should not be more than 15 or 18 inches from the ground, they should be quite flat, and at least 3 inches wide. I constantly see advice to the contrary, but I am convinced that crooked breasts, though sometimes hereditary, are nearly always caused by round perches, while small narrow perches cause curved and crooked toes. Very young chickens will require special treatment, which I will describe later on.

One house and run will not be sufficient if we wish to keep poultry for exhibition; at least two runs will be required for the breeding stock, and also if possible separate runs for the cockerels and pullets. The latter may be dispensed with, and the chickens allowed to run with the old birds, but certainly not to the advantage of the chickens. A very mistaken idea prevails as to the amount of grass run required for Dorkings; if they have five acres they will certainly make use of it, but my own stock birds have never had more than a run of 10 or 12 yards square for each cock and four or five hens, and have always laid and thriven well upon it, while in the show pen their condition has generally been as good as the best; and this is a sure proof to me that no larger run is required, as if fowls are discontented and restless they will rapidly lose that brilliancy of plumage which so surely tells in the prize list.

A single cock for exhibition may be advantageously shut up in a small place if carefully attended to, but the hens become restless in very close confinement, and do not do well under the space I have named. Small shrubs or fir trees planted in the corners will form a pleasant shade in summer, as also will Jerusalem artichokes, which are to be strongly recommended on account of their hardiness and rapid growth. The fences to separate one run from another in my own case are made of hurdling or "wattling" about 3 feet high, with 3-feet wire netting stretched loosely along the top. I have found this quite sufficient to keep the birds in, as Dorkings usually are but poor flyers. If hurdling cannot be obtained, half-inch boards of red deal to the height of 2½ feet, will do as well, but will, of course, be more expensive. Simple wire netting, however small the mesh, is not sufficiently near the ground, as if the cocks can see one another they will fight through it a great deal more than if at liberty, and will never seem to get tired of it, while if at liberty one soon conquers the other, and it is all over except the crowing! If it is determined to try to induce two strange cocks to run together, the only way is to treat them like two schoolboys, and having put bits of cork, or pads, on their spurs, to let them "have it out." If you separate them they are sure to be "at it again" immediately your back is turned.

It will be found very much cheaper in building houses to make them in pairs. The accompanying sketch represents the front view of a pair made in the very simplest way, and with the smallest amount of material, and yet, I think, combining every essential for health and comfort. I have found thin sheet zinc far preferable to felt for roofing, as it is much more durable, and not finding any hold for their claws, the fowls will very soon get tired of flying up on it. A double house of the dimensions described, made of three-quarter inch deal, with zinc roof, should be constructed for about £5, or for less, if, as in my own case, the carpentering be done at home; and really if the planks are already cut out there is but very little carpentering about it, while we shall always think more of our "villa" if made with our own hands. The cracks between the boards had better be covered over with narrow strips of very thin deal, or the boards can be let into one another; but this will be found more expensive.

The little outside sheds are for the fowls to get into in wet and windy weather, and they should always be filled with dry earth or road grit, in which the birds will delight to dust themselves; while being placed outside the roosting houses they will help to keep the birds warm.

CHAPTER IV.

FEEDING.

HAVING built a house and bought birds, the next question will be what to feed them on; and it will be as well here to go through all the different foods, giving the merits and demerits of each.

BARLEY is commonly thought to be the only food that fowls can possibly require, and many wretched birds are shut up in a small yard with nothing but a scanty allowance of this grain, and, to their owner's astonishment, do not pay. Fowls may be truly said to be omnivorous; they will eat and enjoy green food, grain, seeds, insects, worms, and a thousand things we wot not of. They also require access to a heap of lime rubbish, which is to them what salt is to us, besides helping to form the egg shells; and in addition to this they *must have* a supply of small stones to grind it all up with, gravel in the gizzard performing the same office for them that teeth do for us.

It will be at once seen, that however good barley is, it is insufficient by itself to keep birds in good health, and it almost amounts to cruelty to try to do so. For an occasional food it is well enough, but in my opinion it is not equal to WHEAT.

The best wheat is at the present time selling at about 5s. 6d. a bushel, and at this low price is by far the cheapest and best food we can use. I do not approve of tail or offal wheat, as I am sure with all grain the refuse, though low-priced, is by far the dearest in the end. It is very well for a farmer to use such stuff, for it costs him nothing, but to buy such rubbish is a great mistake, more especially in barley, the lightest of which is all husk, and has no kernel. Good wheat, then, is my idea of a food, and one of which the fowls are particularly fond. They lay and do well upon it, and I think no one can err in giving their fowls one feed a day of it.

PEAS and BEANS are not generally used for fowls, but I can strongly recommend them. White peas are the best, and the beans should be cracked in a kibbling machine. They both form a capital occasional food for laying stock, and will bring the birds into splendid feather. They must not be given in excess, nor to chickens which are intended for the table, as they will make the flesh very hard and tough; but as old cocks and hens are generally pretty much this way already, no harm can be done them.

MAIZE, or Indian corn as it is called in England, is a large yellow grain of which fowls are particularly fond. There are two sorts—large and small, the latter the most expensive. Maize is at the present time rather dear, and I do not recommend it except for an occasional change once a week, and then not to the white-feathered birds, or it will most certainly turn their plumage yellow. Some people may laugh at this idea, but such have only to see what cayenne pepper will do for canaries to be at once convinced. Maize is fattening, but it is not a good egg or flesh-former, so is one of the worst foods for growing or laying stock; but it claims one merit—the sparrows cannot possibly swallow it, nor can it be trodden into the mud, and for this latter reason I generally have it used in wet and dirty weather.

OATS will be relished for a change, but they must be sound and heavy or the fowls will not eat them. White oats are preferred, and they should not weigh less than 39 lbs. to the bushel.

RICE I never use, and though apparently cheap, it will be found very dear in the end, as there is no "heart" in it.

BUCKWHEAT, a small dark grain very much like hempseed, is strongly recommended by some, but I could never get my birds to eat it: I have tried it several times, both for old birds and chickens, but it has always ended in waste. I daresay the fowls would eat it if seen, but none are so blind as those that won't see, and my birds seem determined not to see it even when laid on a white plate. A very little HEMP-SEED is not a bad thing in the early part of the year to start the hens laying, but if given in excess it is too forcing, and will cause them to lay eggs without shells. If given in the moulting season it is said to cause the new feathers to come of a darker colour, but as to this I cannot speak from experience, for I don't use above a quart of it in the year.

Grain ground up into MEAL and slaked with water—soft food as it is called—should be given to exhibition birds at least once a day, as, though a little troublesome to manage, it will be very advantageous to the fowls. In cold weather it should be mixed with warm water into a crumbly mass, and if given warm on a winter's morning will greatly promote laying. I always use a zinc bucket and an iron spoon to mix it, first pouring in a little water, and then the meal: if properly prepared it should not be sticky.

GROUND OATS are generally considered to be the best staple food, and my own birds have one feed a day of it all the year round. It must be borne in mind that ground oats are not the same as oatmeal, but the whole grain ground up, husk and all; it is very difficult to obtain good, and I send nearly fifty miles for it, but am convinced that this extra expense is not money thrown away. When good it looks rather like coarse flour, and mixed with water should not show much husk.

BARLEYMEAL by itself is too sticky, and clams to the birds' bills, but if mixed with fine bran (sharps as it is called) it will answer very well; I sometimes mix it with ground oats; but fine bran will do equally well, and is much cheaper.

It will be seen from the above list that there is plenty of choice, and the oftener the diet is varied the better will the birds prosper; but it must be borne in mind that Dorkings fatten more readily than any other breed, and if we wish to keep them in good health they must not be overfed. In winter they may have almost as much as they will eat, but in summer they should be kept very short, especially if they have a grass run.

Some GREEN FOOD is absolutely necessary; if not to be obtained in their yards they should have some given them daily—a mangel wurzel is the best thing I know of, especially as the roots are very cheap, and will keep all through the winter. They should be chopped in half; the fowls will very soon eat the heart out of them.

My own Dorkings are fed twice a-day—early in the morning, and the last thing before roosting time. In addition to this I usually go round the yards at mid-day with a pocketful of corn, and throw them a few grains while I see that they are all as they should be; and it will be well to bear in mind that the "eye of the master makes the horse fat."

One word as to purchasing food. If it is hoped ever to make fowls pay, the grain and meal must not be bought in small quantities, but should be purchased by the quarter or sack. Some friends of my own insist on purchasing it by the gallon, the result being that their birds often have none at all, and what they do have costs twice as much as my own.

Very little need be said on the score of drinking water, save that it should be changed every day, and is best kept in iron vessels, as the latter are not easily broken, and, if a little rusty, will give a chalybeate taste which will be very beneficial. For a number of fowls an iron pig-trough will answer very well, while what are sold as dog-dishes will do very well for a few.

CHAPTER V.

PREPARING FOR EXHIBITION.

THE mysteries of preparing fowls for exhibition seem almost as terrible to the young poultry-fancier as the prospective gridiron to the candidate for masonic honours; yet the latter may not be found quite so hot as expected, nor the former beyond the power of the simplest beginner. I remember well how, when commencing, I was offered all sorts of preparations for putting on the comb and feathers, and without which it was said to be impossible for a bird to win; however, I managed to get on without them then, and shall hope to do so for the future. As some little preparation, however, is required, I will briefly mention all that I think is necessary.

By far the finest thing to get birds into good condition and brilliant plumage is a good grass run, and if this is obtainable little else will be wanted; but if birds are penned up in a small yard some little management will be required to get them into good condition. I have before mentioned what a capital thing peas are, and a few should always be given every other day to exhibition birds; and it will be well to bear in mind, too, that soft food produces soft feather, and that though it will not hurt birds to be hungry, but rather tend to "liven them" up, overfeeding will be certain to give them indigestion, and to turn them black in the comb. It is a common mistake to suppose that Dorkings require to be fattened-up for exhibition. Now-a-days the judges never go by weight. What they look for is a large frame, with plenty of room for putting on flesh. Some breeds look all the better for being fattened, but not Dorkings, which are naturally quite plump enough.

Dorkings stand exhibition worse than any other breed, and if sent from show to show will very soon break down in health, and become as useless for stock as for the show pen. I constantly hear it said that such a bird will "do to breed from," but I don't understand this. If a bird is not moderately perfect it is hopeless to expect perfect chickens from him, and a broken-down constitution is sure to produce degeneracy, if nothing worse. Some birds, though in capital health at home, will mope and look wretched when penned. To such a little toast soaked in strong ale may be given the day before; but it must be borne in mind that all unnatural feeding will surely have to be paid for, and that a little cooked meat with a rusty nail placed in the drinking water should be quite sufficient.

Before going to their first exhibition fowls should always be trained to a pen at home, or else when the judge goes round to look at them they will either get into a corner to hide themselves, or else fly and flap about, and most likely send a lot of dirt and rubbish into the judge's eyes, in which case the latter may be very naturally expected to go on to the next pen. The simplest plan will be to purchase a couple of wire pens and water-tins from one of the well-known contractors who supply our shows, and to fix them in a convenient corner where the young birds may be accustomed to exactly the same place that they will be put into when at exhibitions, and will thus be made to feel quite at home, and to show themselves to the best advantage. I should always advise that the birds be caught at night, as then there will be no struggling or pulling-out of tails, and the birds will quietly submit to their fate. Some management will be required to hold a large bird without damaging his plumage, but a little practice will soon master this. In taking a bird off the perch it will be best to seize him round the wings with both hands, placing the tips of the fingers of one hand under the breastbone to support the body, and in this position you may defy any struggling; but sometimes we can only spare one hand to hold the bird with, the other being required to open the lid of a hamper, administer physic, or wash the dirt from the bird's feet; and here the beginner will find a difficulty, and I must say that until lately I myself was sometimes mastered by a large cock weighing 12 or 13 lbs., but thanks to instructions from Mr. Teebay, I believe I am now a match with one hand for a "regular kicker." The plan is this: place the thumb of one hand over the wing, grasp the thigh firmly with the first two fingers of the same hand, and with the remaining fingers placed under the breastbone you will have the bird fixed as firmly as in a vice. A bird should always be taken out or put into a show pen or hamper head first, otherwise the tail and wing feathers are very liable to be broken. If the fowl is large and the hole small, turn the bird on his side, and he can make no resistance. A fowl should never be caught by the leg, as not only is it a very unworkmanlike plan, but the bird will very likely flap about and hit you two or three smart blows in the face with his wings, besides scratching your hand with the toe nails of his other leg. The best plan is to catch him by the wing, and if the latter is seized close to the body you may hold the bird from the ground by it without the least injury to the bird or yourself.

A basket, as in the drawing, will be found most suitable for sending about exhibition birds. It has no claim to originality, but is here given for the benefit of those who may not know the best shape. The basket must be round inside, and made of light wickerwork, lined with strong calico or thin canvas. The lid should open in the middle, so as to allow plenty of room for the fowls to be put in and taken out without damaging their feathers or comb. The sides need not be filled in with wickerwork, but I have found that only one rim round the middle of the basket is not sufficient to make it firm, so it will be better to have two, as in the engraving. The top must be made of close wickerwork, as, if only made of open work, the railway porters will catch hold of the thin bars to lift it by, and will very soon break it all to pieces; while if the top is made of close wickerwork they will be obliged to catch hold of the handle in the middle.

Preparing for Exhibition. 19

The diameter of a basket for a single cock, or a cock and hen, should be about 24 inches; height 27 inches, to allow the cock to stand up without bending down his comb. For hens 15 inches high will be sufficient. The lid should be firmly tied down with string in two places in case one fastening should become undone. If the birds have to go a very long

Exhibition Basket.

journey a small cabbage or lettuce may be hung up inside the hamper for them to peck at; but anything else put in the hamper will only make their feathers dirty, and corn would be lost. The bottom of the hamper should be covered with clean straw, which should be thrown away when the birds come back, and the hamper put out in the sun to air and freshen for another trip, as a close-smelling hamper with dirty straw is quite sufficient in hot weather to make the birds ill.

The comb, earlobes, and wattles should be sponged over the last thing before starting for the show, first with lukewarm soap and water, and afterwards with cold water. This will make the birds look very fresh and bright, and if they are in good health nothing more will be required; but occasionally it will be better to smear the comb, etc., over with a little fresh butter, and this is all I ever use. I have been recommended salt butter, but though it may make the birds' combs red for the once, it is too strong, and will very likely cause a tender comb to ulcerate. I have also tried oil, vinegar, whisky (!) and everything under the sun, but am sure that simple butter is as good as anything, if not better. The feet should be washed with a brush in warm soap and water in which a very little soda has been placed, and afterwards dried. It will be much the simpler plan to get some-one to hold the bird while these ablutions are going on, but if this cannot be done the bird may, after a little practice, be held between the knees.

CHAPTER VI.

MATING AND SITTING HENS.

DORKING chickens grow and mature faster than any other known variety, and it is for this reason that they are always kept where there is a regular large demand for early chickens of the finest quality. Hamburghs will excel them as layers, but it is very doubtful whether the large size of the Dorking egg does not more than compensate for any deficiency in numbers; and in this as in all other breeds hens vary as layers, but average Dorking hens lay eighty or ninety eggs in the year, besides hatching and rearing a couple of broods of chickens in the most exemplary manner; and notwithstanding all the tall talk to the contrary, I very much doubt if any other breed will do any better. I have heard several complaints this unusually severe winter of people being without eggs; but although I only keep a small number of hens, I have not known what it is to be without them. However, it is on their merits as chickens for the table that Dorkings must rise or fall, and to this I will now confine my remarks.

With prize stock, as with poultry, the desideratum is to produce as large as possible a quantity of the best meat in the shortest possible time—that is, to combine excellence of flesh with early maturity.

What the Shorthorn is to cattle the Dorking is to poultry —that is, not only will this fowl come to maturity sooner than any other, but when matured the principal portion of the meat will be found in the most desirable places—viz., on the breast, wings, and merrythought, instead of on the legs, as is generally the case with Cochins and Brahmas. The advantage this peculiarity gives to the Dorking is very great, and it is for this reason, independently of their white flesh, that they are so much sought after by higglers and dealers; the only fowl which can compete with them in this respect being the French Dorking, the Houdan; but as this variety is a non-sitter it is not generally suited for domestic purposes.

If it is only desired to rear chickens for the table we cannot well hatch them too early in the year, provided they can be kept out of the wet and damp, as the earlier they are the more valuable they will be. But if we hatch out a brood of Dorkings with the hope of finding some prize birds among them, it is very doubtful if anything is gained by beginning too soon.

I never hatch any chickens till February, and often not then. The early birds if not stunted by the cold will doubtless win at the summer shows, and it is for this reason people are so anxious to get them out; but the March and April birds, which have the whole summer to grow in without a check, always make the finest in the end, and it is a common thing to see an April-hatched Dorking winning in November against January and February birds. But not so with Asiatics: it is well known they take nine or ten months to mature, while a Dorking, if pure bred, will be as far advanced at six or seven.

The careful breeder will always like to be certain of having his chickens bred from certain cocks and certain hens, in order to supply on the one side what is wanting on the other, for perfect birds are hardly ever seen except in poultry books. To make sure of this the sexes must be separated in the winter, and the desired birds mated-up again at least a month before we think of setting their eggs.

It is commonly supposed that the strongest chickens are obtained from a cockerel mated with hens, or an old cock mated with pullets. But owing to Dorkings maturing so much earlier than any other breed, this rule, I think, hardly applies to them. I have certainly bred some of my finest chickens from cockerels and pullets running together, and if the latter are early-hatched birds of the previous year they will be at least ten months old and fully matured, and I cannot see that any harm will ensue. It is not advisable to sit the first few eggs of a pullet, as such, besides being small, seldom hatch; but a March or April pullet will, if well fed, be nearly sure to lay in the autumn, and will thus be to all intents and purposes a hen in February or March, and her eggs may be safely trusted.

I have laid great stress on the Dorking maturing so early, but we must also bear in mind that early maturity means early decay. Dorkings are not long-lived birds; the large show cocks seldom get over their fourth winter. I know several birds in a neighbouring farmyard which are for certain twice that age. However, I am not now speaking about the common barndoor Dorking, but the larger bird of the show pen, and I am justified in saying they are short-lived compared with some breeds, as I can hear of no instance of a well-known show cock lasting more than three seasons, while Cochins and Brahmas frequently hold out to five and six. The hens do better, though, and I now know a Dark Dorking hen which has won several prizes, and which is still flourishing and winning, though certainly over eight years old.

It is for this reason—their comparatively short life—that fanciers will not be justified in giving the extraordinarily high prices for Dorkings which Cochins and Brahmas sometimes fetch. We may reasonably expect to have several years' service and many shows out of these latter; but anyone who gives a high price for a Dorking cock, thinking to exhibit and win with him through the season, as is often done in some breeds, will find he is reckoning erroneously, as no Dorking, however carefully attended to, can stand knocking about from show to show, and certainly not for two or three seasons running.

To return to the mating of our birds. If we have an old cock which has not been overshown, and which we desire to use for stock purposes, put him with three or four pullets; but if we have only a nine-months cockerel there will not be the least reason to doubt the probable excellence of his chickens, even though mated with pullets. Besides, we are often compelled to breed from the latter, as it is seldom we can get the old hens to lay soon enough for their eggs to be sat for early chickens. In April and May seven or eight hens may be allowed to one cock, but in the early months it will be better not to allow more than four. When the birds have once been mated-up they should not be separated, as it is hopeless to expect to rear prize chickens if we keep on knocking the parents about from show to show. I should like to see the great winter shows end with December, when we could exhibit our best birds two or three times and afterwards breed from them, but I am afraid there is but very little chance of this coming to pass.

The eggs for sitting should be collected every afternoon, and if they have to be kept should be stored in chaff or bran, and turned every day. I do not believe there is any advantage to be derived from storing the eggs large or small end up, and most certainly a hen if she steals her nest does not do so; but eggs cannot be left for three or four days in one position with impunity. I should not care to sit eggs more than ten days old, and if they have to go a journey they should certainly not be more than two or three. If we wish the chickens to hatch out together the eggs should be as nearly as possible of the same age, as the staler the eggs are the longer will the chickens be in coming out. Dorking eggs are generally very large, and ten or eleven are quite enough for any hen to cover; in fact, nine will be better in the winter months.

The nests may be made of straw or hay either on the ground in a manger, or in any other convenient place which the hen may select. But if we desire to rear a large number of chickens it will be found more convenient to set the hens in boxes or hampers with lids to each, and to take the hens off, and to put them on again every day after feeding—the plan which is generally adopted by keepers—as by this means we shall avoid disappointment caused by the stupidity of the hens in getting on the wrong nests. The hens may be put in pens or coops to feed.

CHAPTER VII.

HATCHING, ETC.

No difficulty will be found in making a broody hen take to a strange nest; the only precautions necessary will be to move her when it is dark, to give her a china nest egg or two, and to keep her shut in upon the new nest till she becomes accustomed to it. It will not be advisable to put the good eggs under her for a couple of days, till she has become thoroughly used to being taken off and put on the nest again by strangers. This should always be done every morning, and the hen put into a coop or other convenient place where she can be easily caught, with food and water for her to eat, and left there for about twenty minutes, and then be replaced on the nest by hand and shut in. If a regular system be adopted it is astonishing how many hens may be kept going in this way with very little trouble.

Two days before the eggs are due to hatch they should be well soused in tepid water when the hen is off the nest, and after being left in the water for two or three minutes they should be replaced, and the hen put on top of them. This will go a long way in preventing the chickens from becoming stuck to the membranous lining of the shell. If fresh, Dorking eggs will hatch on the twentieth day; but if stale they may be a day longer. For this reason the eggs in a sitting should as nearly as possible be of the same age, so as to hatch out altogether; but on no account should the hen be interfered with till the expiration of at least twenty-four hours from the appearance of the first chicken. I consider it waste of time to help chickens out of the shell. I have often done so, but believe it to be far better to let them die at once, as if they are not strong enough to get out of the shell they are pretty sure to die afterwards, and at all events will never make prize chickens. If possible it will be better to sit two hens on the same day, so that if the eggs hatch out badly we may be able to make up one good brood by putting all the chickens under one hen. A little sulphur should always be sprinkled in the nests of the sitting hens, or otherwise they are likely to become infested with vermin.

The hen-coop as illustrated in the accompanying sketch is well worthy of the attention of amateurs. It has been designed by Mr. Henry Lingwood, who for many years has been one of the most successful exhibitors in the poultry fancy. Last year he very kindly let me have one as a pattern, and I found it answer so well that I have asked and received permission from him to have it drawn for general information. It combines every requisite for rearing chickens successfully, and with it no shed or coop-house is required. It also possesses the merit of being vermin-proof when shut up, and this, too, without the ventilation being at all impeded. Owing to an error in shading, the engraving is not so intelligible as I could wish; but I will endeavour to make it a little more simple by explanation. Most hen-coops are entirely inadmissible for outdoor chicken-rearing, for two reasons: one, that the wet drives into the front of the coop; the other, that in heavy rain the wet runs in underneath the sides. In Mr. Lingwood's model coop both these disadvantages are avoided, as the roof projects over the open front, and effectually keeps out the rain; while a wooden tray is made to fit inside (which should be always kept filled with sand or dry earth), which entirely keeps the little chickens out of the wet. The flap-door, which in the wood-cut is seen lying on the ground, hinges at bottom on hooks, and when turned up and secured with a button, fastens the coop up for the night, secure against fox or rat. The top part of the open front is made of small-mesh wire, and should be

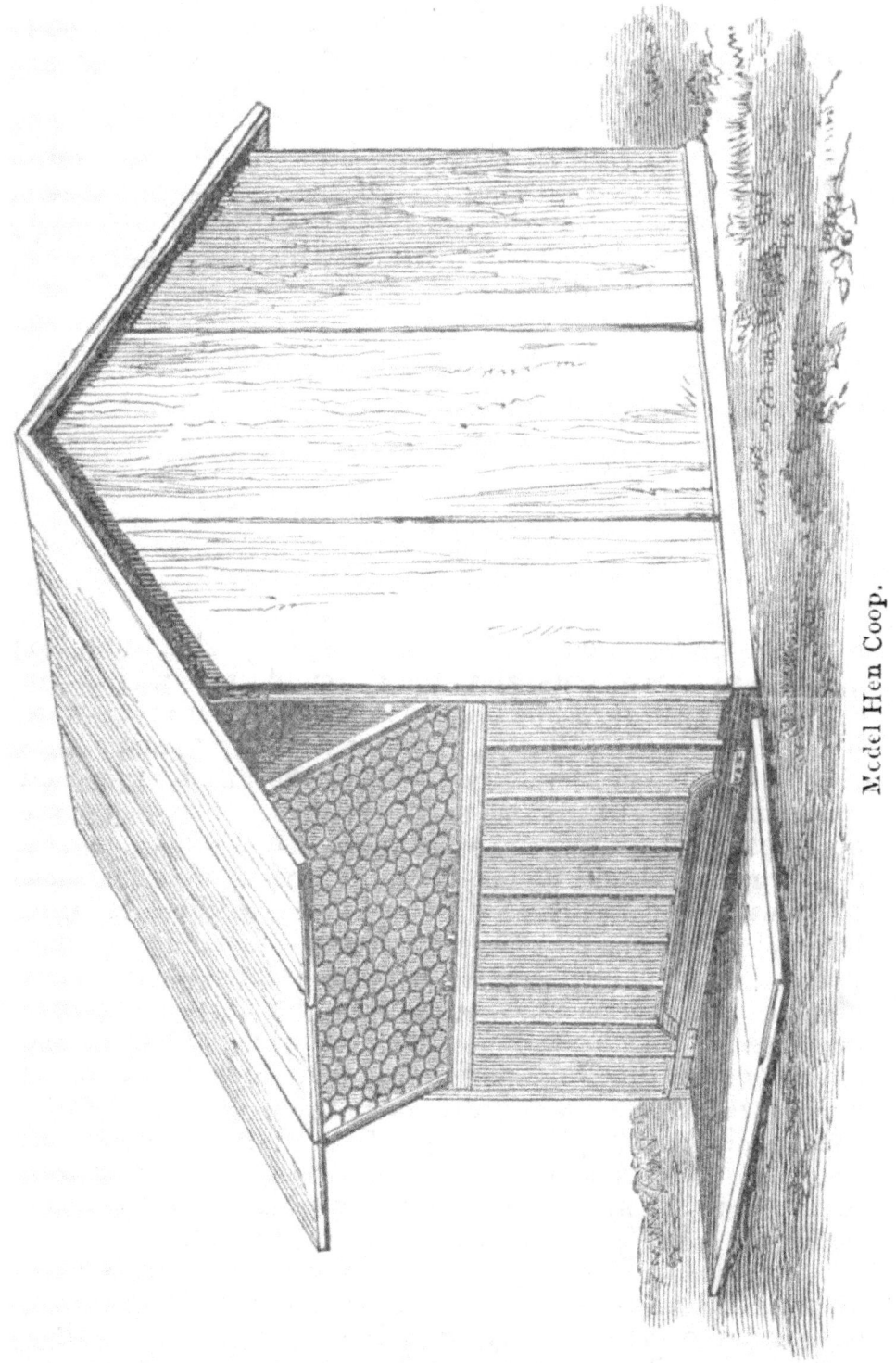

Model Hen Coop.

placed perpendicularly, and not be made, as in the drawing, to lean forward. This, besides giving light and ventilation, will be found convenient for feeding the chickens through on a very inclement morning when it is not desirable to let them run out at once. The best size for them to be made is 2 ft. square at the bottom, and about 2 to 3 feet high. To clear the tray out, the coop must be lifted off; or, the hen may be placed in another coop during cleaning operations.

When twenty-four hours old the chickens may be safely put into such a coop as I have described, or into any other description of coop with a waterproof roof. The hen should be fed and watered, and the chickens too should receive their first feed of bread crumbs softened with milk, canary seed, or some such little delicacy.

CHAPTER VIII.

TO OBTAIN SIZE.

FOR the first fortnight of their lives chickens should be fed at least every two hours, and no amount of care and trouble expended on them will be too much if we wish to rear prize birds. Size is the principal point in a Dorking, and size is only to be obtained by breeding from the finest birds, and by feeding the chickens early and late on the best and most nutritious food. Chickens are awake with the lark; and as it is "the early bird which gathers the worm," we, too, must be up with the lark if we do not wish our chickens to be in the rear when the all-eventful show-day comes. If we happen to have a poultryman who can be depended upon we are indeed lucky, but such are very few and far between, and far better will it be to pay for a little knowledge and become experienced in time than to go on paying high wages to one who very likely knows but little more than yourself, and who will certainly teach you nothing. Two guineas a-week and more are now commonly paid to a first-class poultry manager, and there is but little doubt that if we can afford to keep a poultryman, the best are the cheapest in the end.

After the first day my chickens have whole groats as a staple food, varied with a little hard-boiled egg chopped fine, and ground oats mixed very dry. In a week or ten days they will relish small wheat, and occasionally a little chopped meat may be given them as the egg is discontinued. The great point will be neither to let them eat too much at a time nor leave any, so that, if possible, they may always have an appetite at feeding time. If in the hot weather their appetite seems to leave them, a little gentian bark grated into the meal will be beneficial; but as chickens get older they do not require to be fed so often, and the interval between meals may be gradually extended till, at about four months old, they are fed about three times a-day. If new milk can be obtained they will much enjoy a drink of it the first thing in the morning; but too much coddling will only tend to make them delicate.

It will be a very wise course to weed out the chickens as early as possible, as the fewer there are the better they will thrive. Two or three in every brood may generally be selected for the spit without much trouble, as defective and crooked toes and sooty feet will never get any better. No rules can be laid down for certain, but a little experience will soon teach which chickens to keep.

The chickens should not be allowed to perch at night on narrow or crooked sticks, or they will most certainly become crooked-breasted. Some breeders keep their chickens bedded-down on straw or sawdust till their breastbones are fully formed, but this entails much trouble, and I have not found it essential; besides, some chickens have crooked breasts from the day of their birth, and nothing that we can do will then set them right. Till recently I was of opinion that crooked breastbones could be entirely prevented, but from conversations with some of the most careful breeders, and from my own experience, I am convinced that some cases are hereditary, and, like rickets in children, as long as we breed from such we shall never get rid of it. Undoubtedly birds occasionally receive prizes in spite of crooked breasts, but in my opinion a crooked breast in a Dorking, the table fowl *par excellence*, is quite as bad as or worse than a crooked leg or wry tail. A slight bend in the bone I would pass over, but a bird which has a large hollow in his breastbone I would not have for a gift, and in my opinion he should never receive a prize. If the chickens are given a flat plank of wood to perch on about 6 inches wide, with the sharp edges just planed off so that they cannot perch upon them, they will be obliged to roost on the centre of the board, and if this is not placed too high up we shall see very little of crooked breasts if the parent birds are not defective in this respect.

At ten weeks old the cockerels and pullets should be separated from one another, as by this means a good deal of trouble will be avoided, and the cockerels will not fight, especially if an old cock be left in charge of them. If a cockerel be taken away for three or four days for an exhibition or other purpose he should always be returned to his comrades at night, or a free fight will very likely be the result; but if the bird on awaking finds himself on his old perching place he will forget all about his absence, and will in all probability in the morning run out with the others as usual. This is very important, as with fifteen or twenty cockerels it would be quite impossible to find separate runs for them all.

I have before mentioned how necessary it is to train birds to a pen before sending them to a show, and this is especially the case with hens and pullets. If two hens are caught and placed in a small pen together they will be nearly sure to disagree, and one will peck the other almost to pieces, even though they have been bosom friends before. A good plan will be to first let them run in a small yard together where there are no other birds, and afterwards to put them in a smaller place, and thus to accustom them to one another by degrees. If after this one bird still remains obdurate, and insists on pecking the other, the only way will be to tie the offender by one leg to the side of the pen; but even this will sometimes not effect a cure, and the only way then to proceed will be to make up another pair. The plan of showing two hens in a pen together is now generally given up, though in my opinion it is a far better criterion of who has the best yard of fowls, and I think the yard which exhibits the largest number of noticed birds is more deserving of credit than that which exhibits one, even though this be the first-prize bird.

CHAPTER IX.

DEFORMITIES AND DISEASES.

PRIZE FOWLS, to a certain extent, are reared in an artificial manner; the result is that deformities and diseases are more frequently found in them than in hardily-reared farmyard stock, though the latter are often thought to be more perfect than they really are. Not many farmyard fowls would stand being scanned by a critical eye, while, from what I can find out from farmers, they lose very nearly as many fowls from death and disease as we do among our own more highly reared stock. Anyone thinking of becoming an exhibitor will do well to make himself acquainted with the most common defects, so that he may not be taken in when buying, nor waste entry money in sending out birds which he ought not to expect to win.

A CROOKED BREAST can be easily found out by passing the hand along the breast bone. If the bone is badly curved or indented the specimen should be rejected, as the defect is often hereditary, and will nearly always prevent such a bird from winning in the show pen.

CROOKED BACK, though sometimes apparent, can often only be ascertained by handling. The best way to proceed is to allow the weight of the bird's body to rest on one hand with the legs hanging down, while the other hand is passed down the back. If the spinal column rises in the middle it is called *hump back;* if the spine is twisted, *wry back;* while if a lump is found on one side of the body higher than the other, the *hip bone projects*. The latter deformity often causes the tail to be held on one side, called *wry tail;* but this latter will in a young bird sometimes proceed from weakness and inability to hold the tail upright, but it should always be regarded with suspicion. A *squirrel tail* is when the tail is carried right over the back and almost touching the head. The latter is a great eyesore, and, as well as the preceding malformations, is most surely hereditary.

BUMBLE FOOT is an enlargement or inflammation of the feet or toes. It is not peculiar to Dorkings, but is found in all large breeds, and I have seen several cases of it in the yards of one of the largest and most successful Brahma breeders in the south of England. Sometimes it occurs owing to a thorn or stone being driven into the sole of the foot and occasioning inflammation; but these cases are rare, and I am convinced it is generally an hereditary complaint, like gout in the human subject, and the only way to get rid of it is to ruthlessly blot out every specimen diseased this way, or at least not to breed from them. My old Dorking cock, cup at Birmingham and second at Crystal Palace in 1873, up to his death at nearly four years old, never showed the least trace of bumble foot, nor have I ever found it in any of his chickens; while, on the other hand, a fine bird which I bought cheap to breed from, thinking that a bumble foot would be no detriment for the purpose, threw chickens which showed corns on their *upper* toes when many of the birds were not six months old. The mischief generally begins by a corn growing in the pad of the sole of the foot, and this often proceeds no further. To prevent accidents, though, it will be better to pick this out with a blunt penknife, after first soaking the foot in warm water. Occasionally a little blood may appear, when it will be better to apply lunar caustic to the hole, as is frequently done to stop the bleeding from a leech bite. There will be far less chance of blood flowing if the operation be performed in the moulting season, when the blood flows principally into the newly-forming feathers. If a toe should become enlarged from any cause, accidental or otherwise, the swelling may be reduced, if not of too long standing, by the daily application of tincture of iodine with a camel's-hair brush.

CRAMP occasionally occurs in damp yards. The sufferer should be placed alone in a dry house or pen, and be bedded down in hay or straw, till he recovers the use of his legs. I do not believe that any physic will do good, and patience is all that is required.

DIARRHŒA is occasionally troublesome; it is caused by some irritating substance lodging in the intestine. A large tablespoonful of castor oil is the best remedy, as that will remove the evil. With little chickens a meal or two of boiled rice sprinkled with powdered chalk will prove sufficient in most cases; or small pills may be made up of prepared chalk and chlorodyne, only using sufficient of the latter to consolidate the chalk. Each chicken to receive one twice a-day. This is troublesome, but those who will not endure trouble will not succeed in chicken-rearing.

GAPES.—Little chickens often fall victims to this disease. A small and tenacious worm is developed in the windpipe, which the chicken endeavours to dislodge by repeatedly gaping or opening the mouth, and this symptom gives the name to the disease. A simple remedy is to administer a small pill of pure camphor to each chicken daily. The best preventive is never to allow young chickens to drink stale or impure water, and always to keep camphor in the water.

RELAXED CROP cannot be cured by any medicine, unless perhaps when it first makes its appearance. It is a great eyesore, but when once established cannot be remedied, and so the only plan is to put up with it. If a fowl swallows any hard substance too large to pass into the gizzard, the proper course is to cut open the crop with a penknife and remove the offending matter; the wound should be made no larger than necessary, will bleed very little, and will rapidly heal. I consider slack crop to be hereditary, and by repeatedly breeding from birds with this defect we could doubtless soon produce crops in fowls relatively as large as those of Pouter Pigeons.

ROUP has rightly been called the scourge of the poultry yard. It generally proceeds in the first instance from a bad cold, but when a case is once established in a yard the germs of the disease are carried in the air and breath. The disease generally becomes apparent about the third day of infection; till then it is latent. Preliminary symptoms are sneezing and rattling in the throat, but occasionally one eye will become suddenly closed. When the running at the nose appears the disease is developed, but sometimes birds will run at the nostril without any previous warning. In every large poultry yard there should be an hospital pen in some out-of-the-way corner, where sick birds may be placed on the first appearance of illness, not only for their own comfort, but to prevent contagion. If a roupy bird has once been placed in this pen it should afterwards be kept for roupy birds alone, and should be periodically limewashed. In bad cases of roup, especially in old birds, the eyes close, the head swells, the discharge from the nostril and mouth becomes thick and purulent, and death soon supervenes. If the affected bird is not very valuable it should be killed on the first appearance of the disease, and buried to save the other fowls, as the disease is very contagious. The discharge from the nostrils proceeds from the mucous membrane of the back of the mouth and throat, and the only other disease to which it is analogous is glanders in the horse, the latter being equally infectious. The moment the disease appears Condy's red fluid (principally permanganate of potash) should be added to the drinking water till it becomes of a pinky colour, and chloride of lime should be sprinkled about freely; and both precautions should be adopted through the whole yard. The only internal remedy which I have found effective—and I have tried them all—is a mixture of copaiba and oil of cubebs, in the proportion of four parts of the former to one of the latter, the dose to consist for a large fowl of twenty minims of the mixture made up in the form of a gelatine capsule; or copaiba capsules alone will do. A Dorking cock would require at least three capsules, or about sixty minims, in the day, to effect a speedy cure. In addition to this, the bird's throat, mouth, eyes, and nostrils should be sponged out daily with a strong solution of Condy's fluid, to correct the fœtor of the discharge and to promote more healthy action. I have not known the above remedies to fail, and may claim some little experience, as last autumn I bought ten old farmyard fowls purposely to experiment on. I shut them up in an old wood house at a cottager's with a bird in the most advanced state of roup, her eyes being nearly closed, the discharge from her nostrils most offensive. They had no water to drink, but moist meal twice a day, and on the fourth day eight out of the ten had a running at the nose and the others soon followed, and several of them rapidly became worse. I commenced the treatment above recommended on the eighth day, and before a week was over the whole were cured, including the hen which introduced the disease. The shed where these birds were confined was thatched at top but open at the sides, and consequently the foul air blew away. I think this a better plan than shutting the birds up in a close stuffy pen.

SHELL-LESS EGGS with Dorkings are invariably caused by the hens being too fat: starve the birds for a day, and then feed them principally on vegetable food, to reduce the system.

For all slight indispositions I give a large tablespoonful of castor oil. Fowls will stand as large a dose of aperient medicine as a human being, and the reason why medicine is not generally effective with them is, that not sufficient is given.

Zinc ointment is very useful for cuts or wounds.

CHAPTER X.

THE DARK DORKING.

AMONGST the different breeds of Dorkings, the Dark or Coloured Dorking, on account of its superior size, has always been the favourite. This variety was originally called the Grey Dorking, then the Coloured Dorking, and now is generally known by the name of the Dark Dorking, to distinguish it from the Silver-Grey variety.

The first and most important point in a Dorking is the shape. Most beginners are apt to think that five toes on each foot, and a freedom from leg feathers and crest, are all that are required to make a good bird. Not so; it is the peculiar shape which makes the Dorking—the square, deep, and massive body, and the full chest. The breast bone should be long and deep, to allow plenty of room for putting on flesh, as it is the Dorking's breast which gives it its pre-eminence for the table. The back should be very broad and flat at the shoulders, and should gradually narrow to the hips; the legs should be short and stout, and with plenty of bone, as the leg is a good criterion of frame and capacity for putting on flesh; the thighs should lie close to the body, and not be carried stilt-wise, as in the Cochin. The above points apply equally to all Dorkings.

I will now state what is generally considered requisite in a show bird of the Dark variety; and first for the cock. Head of a good size, eye large and dark in colour, earlobe red,

The Dark Dorking.

wattles long and pendulous, comb either single or double. If the comb is single it must be perfectly upright, quite straight, free from any excrescences at the sides, and evenly serrated with about six or seven sprigs ; if rose-shaped

Dark Dorking Cock.

(double) it should be close and firmly fixed on the head, square in front, flat at the top with the exception of numerous little sprigs, the sides straight and even, and the whole comb narrowing behind into a distinct point. The comb in either case should be of a good size, but without the least tendency to coarseness. Rose-combs in Coloured Dorkings have lately become very scarce in the show pen, and not more than a dozen pens, I think, have been seen at the Crystal Palace Show for the last five years. They are evidently not popular, the reasons doubtless being that they are more difficult to find in perfection, that they invariably become coarse and ungainly in the second year, and that they are more open to being tampered with than the single combs; this latter in my opinion being a serious objection.

Rose and single-combed birds may be bred together, and a proportion of both will be found in the chickens; but this plan should be avoided if possible, as coarse and ugly combs are sure to be the result, and neither comb will become fixed in the strain. In a pen of birds for exhibition of course the combs must match; if a rose-combed cock and a single-combed hen were shown together certain disqualification would be the result. Sometimes we find a bird with a comb something between the two—that is, a single comb in the front part expanding into two combs, as it were, in the middle, and joining again at the back, so as to form a large hollow. These "cup" combs as they are called will not now pass muster, and are only fit for the farmyard. The neck hackle should be very full, and should fall naturally, the feathers not twisting over one another or rising in a hump behind the neck; tail very large and full, with the feathers unusually broad and arching, and the saddle hackles very plentiful; legs short and perfectly straight; toes, five on each foot, the fourth and fifth or hind toes being very distinct, and each growing separately from the leg, the fifth toes turning upwards; the front toes long, straight, and well spread, and the spurs set well inside the leg, in fact almost pointing at one another.

Formerly the colour of a Coloured Dorking cock was immaterial, but of late years, since a separate class has been made for Silver-Greys, this is not the case. The neck hackle should be clouded or striped with black, and not be white, as in the Silver-Grey. The saddle, too, should be clouded, and the shoulders of a darkish tinge. The accompanying illustration, drawn from a cockerel of my own breeding, gives a very good idea of the colour required in a Dorking cock to match the Dark hens, which are now so fashionable. A cock will pass muster if he is not so darkly striped in the neck hackle; but a Coloured Dorking Cock with almost white hackle and shoulders is not the proper mate either for the show pen or breeding for a Dark hen, any more than a Silver Duckwing Game cock is a match for a Brown Red Game hen. To a critical eye one is quite as objectionable as the other.

Occasionally we see cocks showing a good deal of red or chestnut on the shoulders. These will pass with a Dark hen and often breed very good pullets, but they are not so suitable in my opinion as cocks only showing black and grey. It is commonly thought that white in the breast or tail is a bad point. This is a mistake, and arises from confusing the colours of the Dark and Silver-Grey varieties. A little white in the breast of a Dark cock or on the thighs is no detriment, nor is white in the tail; in fact, I have never seen a really good Dark cock without white in the sickle feather,—a "gay" sickle as it is termed. My first-prize cock at the Palace and first cockerel at Birmingham this year both had white in the sickles, as also had my second Palace and cup Birmingham cock the year before, and I have yet to learn that it is any disadvantage. Of course a very white breast or tail is a disadvantage; but it is no disqualification if a bird is good in other respects, and often appears after a moult in a bird which was previously quite free from it.

I have said a good deal about the white in the tail, as the only results of rejecting gay-sickled birds would be to reduce the size of Dorkings by making the choice of show specimens much more difficult, or else to introduce the plucking-and-trimming system, which is such a curse to many other breeds. The colour of the legs and feet should be white; the colour of the flesh and skin of a fowl is always of the same tinge and colour as the feet. So in the Dorking, which is essentially the table fowl, it is very important that this should be white.

I have now mentioned most of the necessary points, and to assist beginners will now state the most common defects. Many Dorkings have undoubtedly been crossed some time with other breeds, such as Cochins or Malays; but such crosses can nearly always be detected. Long legs and deficiency of chest are certain signs of a cross, as also are yellow beaks and toe nails. A small tail, too, should be looked upon with suspicion, as a large and flowing tail is inseparable from a pure-bred Dorking. The comb in the cock sometimes falls to one side and lops behind; both these are signs of lack of condition and health, and are ruinous in the show pen, as also is a bad twist or bend in the comb, though a slight turn or "thumb mark" in the front of the comb is not so important as in a Spanish cock. A crooked leg is not at all an uncommon defect, and should be carefully looked to in buying a bird; but the feet, as a rule, are where the most attention is required, as not only are corns found under the feet, but the hind toes are frequently badly set on. The big toes often curve inwards and the front toes are sometimes set too closely together instead of being well spread. These two last are bad defects. Spurs are found outside, too high up, or at the back of the leg; these blemishes are fast becoming common, and require to be looked to. A decided dark tinge or "sootiness" in the feet is certain disqualification, as also are feathered legs.

CHAPTER XI.

THE DARK DORKING HEN.

THE DARK DORKING HEN should be similar in shape to the cock, with a square, broad, and lengthy body set on short legs. The comb, if single, should be large and falling to one side, a small upright comb being a great eyesore. By some it was thought impossible to breed cocks with upright combs from hens whose combs fell over, but this fallacy was contradicted by Mr. Hewitt in the *Journal of Horticulture* for June, 1867, and I copy his notes, which will be interesting and instructive to many. Mr. Hewitt says:—"In giving my opinion as to the formation of the comb of a Grey Dorking hen, I quite agree that the comb should be of moderate size, well serrated, and hanging over the face on one side. The theory propounded of breeding from what is commonly known as a 'prick-combed' Grey Dorking hen has been again and again attempted, but I never yet knew a single individual who persisted in the experiment a second year, for disappointment was the invariable issue, and the chickens thus produced proved utterly useless for exhibition. Anyone purposing to breed Dorkings I would strongly advise to mate together a cock having a perfectly upright comb with hens whose combs fold and then turn over the face, it matters not on which side. He will then find, if well-bred stock birds, that all his chickens will have combs exactly of the same formation as the parent birds. In Spanish fowls the same rule again holds good. It is only great age or want of health and condition that will cause the combs of either Spanish or Dorking cocks to fall over if they are truly-bred birds. In the latter case it very frequently happens that restored constitution causes the comb to become again as exact and firmly fixed as ever. The different formation of the comb is in the breeds referred to simply characteristic of sex, as in the mane of a lion or the antlers of a buck."

The Dark Dorking Hen.

As regards the colour of a Dark Dorking hen a good deal of latitude is allowed, provided the general appearance of the bird is of a rich dark colour. Whatever a few may say to the contrary, it is impossible to ignore the fact that the old

Dark Dorking Hen.

Brown and Light Grey birds do not now meet with favour either from judges or exhibitors; partly owing to the fact that the Dark hens handle the best from being more tightly feathered, and partly because they present a more attractive appearance to the eye. My taste inclines in the hen to a jet-black neck hackle and tail, the back and wings being of a dark grey colour, and each feather spangled at the ends with a darker marking, the shaft of each feather being distinctly white.

This latter feature is invariably found in all good Dorking hens, and adds very much to the general appearance by contrasting with the dark feathers. With a black hackle I like to see a salmon-coloured breast, but the hackle is often striped with white, and the breast colour may be of any tinge from light salmon to dark chocolate, as long as it does not present a washed-out appearance.

Colour is far more difficult to obtain than size. The latter may be obtained by good and judicious feeding and breeding from large-sized birds not too nearly related; but nothing will bring colour unless it is in the strain. Beginners are apt to think that all they have to do is to buy a couple of good hens and a highly-commended cockerel at a large show, and to breed from them; but this will nine times out of ten end in failure, unless the cock comes from a good strain; as, no matter how dark and good the hens are, if the cock be not also bred from dark-coloured birds it is impossible to say what the chickens will turn out. Before buying a cockerel to breed from I like to see what the stock hens of the yard are like. It is very bad policy to take the show birds as a sample, as these have very likely been bred by someone else. Though the darker-coloured cocks as a rule are most likely to breed dark chickens, this is by no means a certainty, and we shall never know what our chickens are to be like unless we know for certain that the cock comes of a good strain.

The principal defects to avoid in a Dark Dorking hen, in addition to what I have alluded to when speaking of the cock, are upright comb and "sooty" feet, the latter being much commoner in the hens than in the cocks, as also are white earlobes. A red or rusty colour on the wings is also to be avoided.

Dorking eggs are white and unusually large. The chickens when hatched are very pretty, and should be uniform in appearance, with a broad dark band down the centre of the back, and with two narrow white stripes on each side of it. As a rule the little chickens with the broad band of colour down the back of a sound dark tinge are more likely to make dark chickens; and even from the very best birds it will be futile to expect more than two or three in a brood to reach the required standard of shape, size, comb, colour, and feet, while not one bird in a thousand is perfect in every respect.

The best of rearing a number of Dorking chickens is, that

the "wasters"—birds which can never pass muster—are at the very best for the table when at about ten weeks old, and should then weigh from 2½ to 3 lbs. each. I constantly weigh my chickens, and find that cockerels at two months old should weigh from 2¼ to 2½ lbs. if they are ever to be the heroes of a show pen. At four and a half months old 7 lbs. is a fair weight, and at six months many weigh 9 lbs., and some 10 lbs. Old cocks will weigh from 9½ to 14 lbs., and hens from 8 lbs. to 11 lbs. I of course allude to show specimens; but a cock weighing 10 lbs., if dark and perfect, is likely to win many prizes, as is a hen of 8½ lbs. These weights are mentioned only for the sake of comparison, and birds may generally be made 2 or 3 lbs. heavier by fattening; but this would be worse than useless, as birds now are never weighed, and it is frame and bone which carry the day.

It is impossible to give a standard of points of any real use, as judges will always differ as to the amount to be allowed for condition, nor will they agree as to the relative value of shape, size, and colour; but the following scale gives my own idea on the subject. Allowing one hundred marks for a bird perfect in size and every fancy point, I would deduct the following marks for the following defects:—

Want of shape	from 5 to 25	points.
Want of size	„ 5 „ 25	„
Bad colour	„ 5 „ 15	„
Badly-formed feet and toes	„ 2 „ 15	„
Defective comb in hen	„ 1 „ 5	„
Defective comb in cock	„ 2 „ 10	„
Sooty feet	„ 2 „ 15	„
Slight bumble foot	„ 10 „ 20	„
Spur rather outside (cock)	„ 15 „ 20	„
Absence of white shaft from the hen's feathers	„ 5 „ 10	„
Want of condition	„ 15 „ 25	„
White ear lobe	„ 1 „ 3	„

All deformities, bad bumble feet, spurs RIGHT outside, and very dark feet are fatal disqualifications, and I think badly-crooked breastbones should be so too.

CHAPTER XII.

SILVER-GREY DORKINGS.

THE SILVER-GREY DORKING is really a beautiful variety, and combines the useful points of the Dark Dorking with most attractive plumage. It is to be regretted that this breed is so little patronised, as I am sure no other purely fancy breed can compete with it where fowls have to find their own food;

Silver-Grey Dorking Cock.

nor do they ever look to greater advantage than when working hard for their living in a hedgerow or homestead.

The shape of the *Silver-Grey Dorking cock* differs in no respect from that of the Dark Dorking: the difference is solely in the colour; and as this is a fancy breed, colour is of greater importance than in the Dark variety. The comb of the Dark Dorking may be either double or single, but the comb of the Silver-Grey should be single. This is well recognised, though I have never seen it mentioned. The plumage of the Silver-Grey cock must be solely of black and white feathers, any red or parti-coloured feathers being a disqualification. The neck hackle should be a clear white, but a slight pencilling of black down the centre of each feather is allowed, provided it does not produce a dark appearance. The back, shoulders, saddle, and saddle hackles should also be of a pure white, while the breast, tail, and thighs should be black, any white in these parts being a disqualification, except on the thighs, where a little grizzling in an old bird will pass. The black wingbow should be cleanly cut, and this makes a pleasing contrast to the white of the remainder of the wing.

The plumage of the *Silver-Grey hen* is exceedingly pretty, the whole of the body, feather, and wings being of a clear silver-grey colour, while the breast is a rich robin red. As in the Dark Dorking, the shaft of each feather should be distinctly white; only, in the Silver-Grey variety, this should not extend to the wings and shoulders, or it rather spoils the general appearance. The neck hackle should be white, with a narrow and distinct black stripe down each feather.

Although good Silver-Grey hens are by no means easy to breed, the cocks are far more difficult to obtain perfect; and I often think when I see judges withholding prizes in the Silver-Grey classes while they give prizes to Mealy Cochins and White-throated Brahmas, that they do not make due allowance for the very great difficulties which Silver-Grey Dorking breeders have to contend against.

By far the worst point in a Silver-Grey cock is a yellow hackle or saddle: it spoils the whole beauty of the breed. The smallest speck of white in the breast of the cock or in the tail is considered a disqualification, though I would far rather see this than a yellow hackle. The back and shoulders must be free from red or chestnut feathers: this is positive. The earlobes should be red.

The defect most difficult to avoid in breeding the Silver-Grey hens is a rustiness or dulness of the plumage, especially on the wings. Birds showing this defect should be avoided either for breeding or the show pen, as cocks bred from them would be sure to breed rusty-winged pullets. I should not discard a hen altogether if large and only slightly ruddy on the wing, as this will wear away a little with age; but a white breast or tail in a cock will infallibly become worse with the moult, though in the cockerels the hackles become lighter up to about seven months old.

The Silver-Grey Dorking seldom or never attains the size of its darker brother; but cocks of 10 lbs. in any breed are quite large enough for all useful purposes, and I should be glad to see the judges look more to feather, shape, comb, and feet than to mere size. "Sooty feet" are not uncommon, and they must be carefully avoided. I have seen it remarked that the "sooty" feet now so common in Dark Dorkings have been caused by the overbreeding for dark feather; but if so, how is it that dark feet are quite as common with the Silver-Greys?

To breed Silver-Greys perfect birds must be obtained, though they be small; and the importance of obtaining cocks of a good strain cannot be over-rated. Birds of different strains sometimes produce very badly-coloured chickens if the two strains do not amalgamate or "nick," as it is called; and in commencing a strain it will be far better to procure the cockerels and pullets from the same yard, and afterwards to buy hens of the correct colour to cross with, as by this means size may be kept up with less danger of the whole yard being spoilt.

To obtain early chickens it will be necessary to breed from cockerels, as I have found from painful experience that two-year-old cocks are not to be trusted till the warm weather appears.

The RED-SPECKLED DORKING is now almost obsolete, having given way to the more fashionable Dark and Silver-Grey varieties. Specimens may still be seen in the farmyards of Surrey and Sussex, but of no great excellence.

The principal peculiarity is found in the hens, which are of a brown or chocolate colour, spangled all over with distinct white spots of the size of a large pea, giving them the appearance of having been out in a fall of snow. This colour is, I think, more curious than attractive. Cocks to match such hens should have a good deal of red and chestnut on the back and shoulders, their breasts should be spotted with white or red, and white in the tail is not objectionable. I believe they are a hardy breed and very readily fatten.

CHAPTER XIII.

WHITE DORKINGS.

These are great favourites in all country places, their snowy plumage and coral combs never showing to greater advantage then when contrasting with the green of a country meadow. They do not do well, however, for a large town, as their bright plumage very soon becomes soiled.

A White Dorking's comb must be rose-shaped (double), firmly fixed on the head, and without any tendency to lop on one side. A single comb is a disqualification, but is now rarely or never seen. The ear-lobe, as in all Dorkings, should be red, and the feet white. The legs and feet are frequently found of a pinkish tinge, but white is the proper colour for the leg of a Dorking. The colour of the feathers should be a bright spotless white, and it is here where the difficulty occurs, as so many birds show a yellow tinge in their plumage. Many birds will moult-out white, but if exposed to the influence of the sun will rapidly turn yellow. The cocks especially will become yellow on the hackle and saddle; their feathers, being brighter in colour, are more prone to absorb the sun's rays. Some cocks will keep much purer in colour than others, and, for breeding, we must be careful to select birds with this peculiarity in addition to their possessing a good comb and all the Dorking characteristics. Yellow beaks and toe nails should be avoided, as clearly showing a cross.

White Dorkings are not very great favourites with the show community, owing to the great difficulty found in keeping them white and clean enough for the show pen. As soon as the sun attains any power the cocks to be shown are better kept out of the sun, or their plumage will soon become tanned; but with every care, white birds occasionally require washing before exhibition.

The simplest plan for *washing fowls* is to fill a large crockery-pan or tin vessel with about 8 inches of warm soft water, to souse the bird well in it, and wash thoroughly with a soft sponge and white curd soap till the dirt is all out; then pour in fresh water and wash out the soap, and remember that as long as any soap remains the feathers will never web properly. After rinsing thoroughly, squeeze out as much water as possible. The bird should be lightly and quickly dried with a soft Turkish towel, and then be placed on some clean straw to dry. Each bird should be placed in a separate basket before a bright fire or in the sun, and in about twelve hours they should be quite dry and beautifully white. Sour milk is a capital thing to remove sunburn—that is, if the yellow tinge is caused by the sun and is not natural to the feathers.

White fowls should not be allowed to eat maize, or, as I have before stated, it will be very difficult to keep the plumage white. The best dust-bath for them is a heap of white drift sand or road grit.

The eggs are generally white in colour, but Mr. Cresswell informs me that some birds of his well-known strain lay pink eggs, and these must certainly look very pretty on the breakfast-table. The chicks when hatched out should be of a yellowish-white colour.

CHAPTER XIV.

CUCKOO DORKINGS.

DURING the last two or three years these have advanced much in public favour. They have now a class allotted them at the Crystal Palace and at two or three of the other principal shows, and as several fanciers are taking them up we shall probably see more of them in future.

They are called Cuckoo Dorkings on account of their feathers being somewhat similar to those found on the breast of our summer visitor the cuckoo; but they are also called Blue Dorkings. The term "speckled Dorking" is sometimes applied to them erroneously. They appear of a blue or dark blue colour, but on examination it will be found that every feather is evenly marked with broad bars of dark blue or dark grey, on a whitish-grey or light blue ground.

In breeding these birds many chickens will come so darkly marked as to be almost black; these must be discarded, as we must not forget the original blue-mottled breast of the cuckoo, to which we have to breed. They are also very liable to sooty feet. The chief point to look to in exhibiting them is, that the cock and hen are of the same shade of colour, and that both are evenly marked all over. Rose combs are correct in this variety. The principal faults in the cocks are yellow hackles and saddles. They are too, commonly found with white in the sickle feathers. A per-

Cuckoo Dorkings.

fect cock with a Cuckoo tail is a rarity; and as size is one of the principal points in a Dorking, it will not do to sacrifice an otherwise good bird for this slight blemish.

The hens are really wonderful layers,—in fact, quite equal

Cuckoo Dorking Cock.

to Hamburghs; while, though for the table they do not reach the size of the Dark Dorking, their full breasts and juicy flesh, combined with very early maturity, render them most acceptable in the larder. They are decidedly a hardy breed.

In conclusion, I will mention what I consider are the proper classes to be given to Dorkings in a poultry-show schedule. If only two classes are given they should be for Dark Dorkings and for "any other variety." If three classes are given they should be for Darks, Silver-Greys, and for Whites or Cuckoos. Many committees do not receive as many entries as they should do by classing the Silver-Greys with the Dark birds; the result being that the Silver-Grey birds are never entered in the class, for their defeat is certain. If there is only one class for Dorkings it is very rare to see anything but Darks, as a consequence of which the Committee lose the Silver-Grey, White, and Cuckoo entries, which if three prizes were given would be sure to make good classes.

INDEX.

Advantages of Dorkings, 5
Approval, Buying on, 9

Basket for Travelling, 19
Bumble Foot, 12, 29
Buying Birds, 8
Broody Hens, 23

Catching Birds, 18
Classes for Dorkings, 46
Cold, Catching, 30
Colour, 34, 37
Coloured Dorking, 32
Combs, 34, 36
Condition, 17
Coops, 25
Cramp, 30
Crop, 30
Crooked Breasts, 12, 27
Cuckoo Dorkings, 44

Dark Dorkings, 32
Defects, 35, 38, 43
Deformities, 28
Diarrhœa, 30
Diseases, 28
Dust Baths, 44
Dusting Sheds, 13

Ear Lobe, 32
Eggs, Colour of, 44
  „ Storing, 21, 23
Exhibition, Preparing for, 16, 17, 28

Feet, 34
Fighting, 13
Food, 13

Gapes, 30

Hackle, 34
Hamper, 19
Hewitt, Mr., on Combs, 36
Houses for Poultry, 13

Length of Life, 22

Moulting, 15

Nests, 23

Perches, 11, 12, 27
Poultryman, 26
Prices of Birds, 5, 9
Profits, 5

Railway Journeys, 9
Red Speckled Dorking, 42
Roup, 30
Run, Amount Required, 12

Scale of Points, 39
Selling Classes, 8
Shape, 32
Sickle Feathers, 35
Silver-Grey Dorking, 40
Size, 26
Sooty Feet, 36
Speckled Dorking, 42
Spurs, 36

Toes, 34

Washing, 43
Wasters, 9
Water for Drinking, 16
Weights of Birds, 39
When to Feed, 16
White Dorkings, 43

www.ingramcontent.com/pod-product-compliance
Lightning Source LLC
Chambersburg PA
CBHW060005230526
45472CB00008B/1954